A Special Book for
Treasured Memories

To ______________________________

From ____________________________

Published in 1993
by Ebury Press Stationery
an imprint of Random House UK Ltd
Random House
20 Vauxhall Bridge Road
London SW1V 2SA

Illustrations courtesy of
The Amoret Tanner Collection

Border from wallpaper
design courtesy of
Today Interiors
London SW3

Set in Caslon
by FMT Graphics Ltd
London SE1
Printed in Singapore

Designed by
Sara Robin
Picture research by
Philippa Lewis
Editorial research by
Caroline Taggart

ISBN 0 09 177548 5

LITTLE GEMS

Nature's Beauty

A KEEPSAKE BOOK

EBURY PRESS STATIONERY

THE
"BOY'S OWN" CALENDAR 1885
JANUARY.
FEBRUARY.
MARCH.
APRIL.
MAY.
JUNE.
WINTER
SPRING
SUMMER
AUTUMN
JULY.
AUGUST.
SEPTEMBER.
OCTOBER.
NOVEMBER.
DECEMBER.
FRED. MILLER

THEN THE LITTLE HIAWATHA
LEARNED OF EVERY BIRD ITS LANGUAGE
LEARNED THEIR NAMES AND ALL THEIR SECRETS
HOW THEY BUILT THEIR NESTS IN SUMMER,
WHERE THEY HID THEMSELVES IN WINTER,
TALKED WITH THEM WHENE'ER HE MET THEM,
CALLED THEM 'HIAWATHA'S CHICKENS'.

Henry Wadsworth Longfellow
The Song of Hiawatha

I'D BE A BUTTERFLY BORN IN A BOWER,
WHERE ROSES AND LILIES AND VIOLETS MEET.

Thomas Haynes Bayly
I'd be a Butterfly

ART THOU THE BIRD THAT MAN LOVES BEST,
THE PIOUS BIRD WITH THE SCARLET BREAST,
OUR LITTLE ENGLISH ROBIN?

William Wordsworth
The Redbreast Chasing the Butterfly

THE HUNDRED PERFUMES OF THE LITTLE FLOWER-GARDEN BENEATH SCENTED THE AIR AROUND; THE DEEP-GREEN MEADOWS SHONE IN THE MORNING DEW THAT GLISTENED ON EVERY LEAF AS IT TREMBLED IN THE GENTLE AIR; AND THE BIRDS SANG AS IF EVERY SPARKLING DROP WERE A FOUNTAIN OF INSPIRATION TO THEM.

Charles Dickens
The Pickwick Papers

MY LOVE IS LIKE A RED, RED ROSE
THAT'S NEWLY SPRUNG IN JUNE;
MY LOVE IS LIKE THE MELODY
THAT'S SWEETLY PLAYED IN TUNE.

Robert Burns
A Red, Red Rose

CHRISTMAS IS HERE:
WINDS WHISTLE SHRILL,
ICY AND CHILL,
LITTLE CARE WE.

William Makepeace Thackeray
The Mahogany Tree

Wishing a happy new year.
Let us model our spirit to chance and change,
Let us lesson our spirit to hope and range
Through pleasures to come-through years unknown
But never forget the time that's flown!
BARRY CORNWALL.

WHEN ALL THE WORLD IS YOUNG, LAD,
AND ALL THE TREES ARE GREEN;
AND EVERY GOOSE A SWAN, LAD,
AND EVERY LASS A QUEEN.

Charles Kingsley
Songs from the Water Babies

But the rose was awake all night for your sake,
knowing your promise to me;
the lilies and roses were all awake,
they sigh'd for the dawn and me.

Alfred, Lord Tennyson
Maud

BACK ON BUDDING BOUGHS
COME BIRDS, TO COURT AND PAIR,
WHOSE RIVAL AMOROUS VOWS
AMAZE THE SCENTED AIR.

Robert Bridges
Spring

Winter

He did not hate the Winter now, for he knew that it was merely the Spring asleep, and that the flowers were resting.

Oscar Wilde
The Selfish Giant

The red rose cries, 'She is near, she is near;'
And the white rose weeps, 'She is late;'
The larkspur listens, 'I hear, I hear;'
And the lily whispers, 'I wait.'

Alfred, Lord Tennyson
Maud

CHRISTMAS
Sing, Robin, Sing,
though the Snow is
on thy wing,
Sing, Robin, Sing
Sing, little wren,
Sing happiness
to hearts of men
Winter shall not
hush your lay,
Christ was born
for man to-day
FRED E WEATHERLY.
COPYRIGHT
H. BRIGHT.

So the Nightingale sang to the Oak-tree, and her voice was like water bubbling from a silver jar.

Oscar Wilde
The Nightingale and the Rose

THAT WHICH WE CALL A ROSE,
BY ANY OTHER NAME WOULD SMELL AS SWEET.

William Shakespeare
Romeo and Juliet

WE'LL LIVE,
AND PRAY, AND SING, AND TELL OLD TALES, AND LAUGH
AT GILDED BUTTERFLIES.

William Shakespeare
King Lear

Wishing you the Compliments of the Season

GATHER YE ROSEBUDS WHILE YE MAY,
OLD TIME IS STILL A-FLYING:
AND THIS SAME FLOWER THAT SMILES TO-DAY,
TO-MORROW WILL BE DYING.

Robert Herrick
To Virgins, to Make Much of Time

HER CHARIOT IS AN EMPTY HAZEL-NUT,
MADE BY THE JOINER SQUIRREL OR OLD GRUB,
TIME OUT O' MIND THE FAIRIES' COACH-MAKERS.

William Shakespeare
Romeo and Juliet

When the New Year opens
may there be
Many good things in store for thee

COME, HEAR THE WOODLAND LINNET,
HOW SWEET HIS MUSIC! ON MY LIFE,
THERE'S MORE OF WISDOM IN IT.

William Wordsworth
The Tables Turned

FROM THE TIME OF THE ARRIVAL OF THE FIRST SWALLOW THE FLOWERS TAKE HEART; THE FEW AND SCANTY PLANTS THAT HAD BRAVED THE EARLIER COLD ARE SUCCEEDED BY A CONSTANTLY ENLARGING LIST, TILL THE BANKS AND LANES ARE FULL OF THEM.

Richard Jefferies
Nature on the Roof

May your Easter be Happy

SOON WILL THE HIGH MIDSUMMER POMPS COME ON,
SOON WILL THE MUSK CARNATIONS BREAK AND SWELL,
SOON SHALL WE HAVE GOLD-DUSTED SNAPDRAGON.

Matthew Arnold
Thrysis

Christmas comes! Dear friend to thee
Ever happy may it be!

AT CHRISTMAS I NO MORE DESIRE A ROSE
THAN WISH A SNOW IN MAY'S NEW-FANGLED MIRTH;
BUT LIKE OF EACH THING THAT IN SEASON GROWS.

William Shakespeare
Love's Labour's Lost

SILENCE ACCOMPANIED, FOR BEAST AND BIRD,
THEY TO THEIR GRASSY COUCH,
THESE TO THEIR NESTS,
WERE SLUNK, ALL BUT THE WAKEFUL NIGHTINGALE;
SHE ALL NIGHT LONG HER AMOROUS DESCANT SANG.

John Milton
Paradise Lost

Our Bantams

WONDROUS THE GODS,
MORE WONDROUS ARE THE MEN,
MORE WONDROUS, WONDROUS STILL,
THE COCK AND HEN.

William Blake
Miscellaneous Epigrams

NEW YEAR WELCOME.
Clasp hands, dear ones,
when the New Year comes;
Let all mean dissension
cease,
The Old Year has past,
this may be our last;
Let its heir succeed
in peace.
(COPYRIGHT)
E. H
ENT. STA. HALL

WHITE BUTTERFLIES IN THE AIR;
WHITE DAISIES PRANK THE GROUND:
THE CHERRY AND HOARY PEAR
SCATTER THEIR SNOW AROUND.

Robert Bridges
Spring Goeth All in White

ALL THINGS ARE SYMBOLS: THE EXTERNAL SHOWS
OF NATURE HAVE THEIR IMAGE IN THE MIND,
AS FLOWERS AND FRUITS AND FALLING OF THE LEAVES.

Henry Wadsworth Longfellow
The Harvest Moon

O COULD I FLY, I'D FLY WITH THEE!
WE'D MAKE, WITH JOYFUL WING,
OUR ANNUAL VISIT O'ER THE GLOBE,
COMPANIONS OF THE SPRING.

Michael Bruce
To the Cuckoo

I REMEMBER, I REMEMBER,
THE HOUSE WHERE I WAS BORN,
THE LITTLE WINDOW WHERE THE SUN
CAME PEEPING IN AT MORN.

Thomas Hood
I remember

CHRIST-
-MAS
Emblem of grace and joy,
I send this blossom fair;
So may thy life alway
Be sweet with fragrance rare!

Rough winds do shake the darling buds of May
and summer's lease hath all too short a date.

William Shakespeare
Sonnets

Summer

THE EXCEEDING BEAUTY OF THE EARTH,
IN HER SPLENDOUR OF LIFE, YIELDS
A NEW THOUGHT WITH EVERY PETAL.

Richard Jefferies
The Pageant of Summer

THAT'S THE WISE THRUSH,
HE SINGS EACH SONG TWICE OVER,
LEAST YOU SHOULD THINK HE
NEVER COULD RECAPTURE
THE FIRST FINE CARELESS RAPTURE.

Robert Browning
Home-thoughts, from Abroad